EXPLORE WORLD

PHYSICAL SCIENCE

Robots

MICHÈLE DUFRESNE

TABLE OF CONTENTS

What Is a Robot?..2
What Can a Robot Do?..6
Robots and Operations.......................................12
Glossary...16

PIONEER VALLEY EDUCATIONAL PRESS, INC

WHAT IS A ROBOT?

Here is a robot.

A robot can be big or small.

Some robots look like people and some do not.

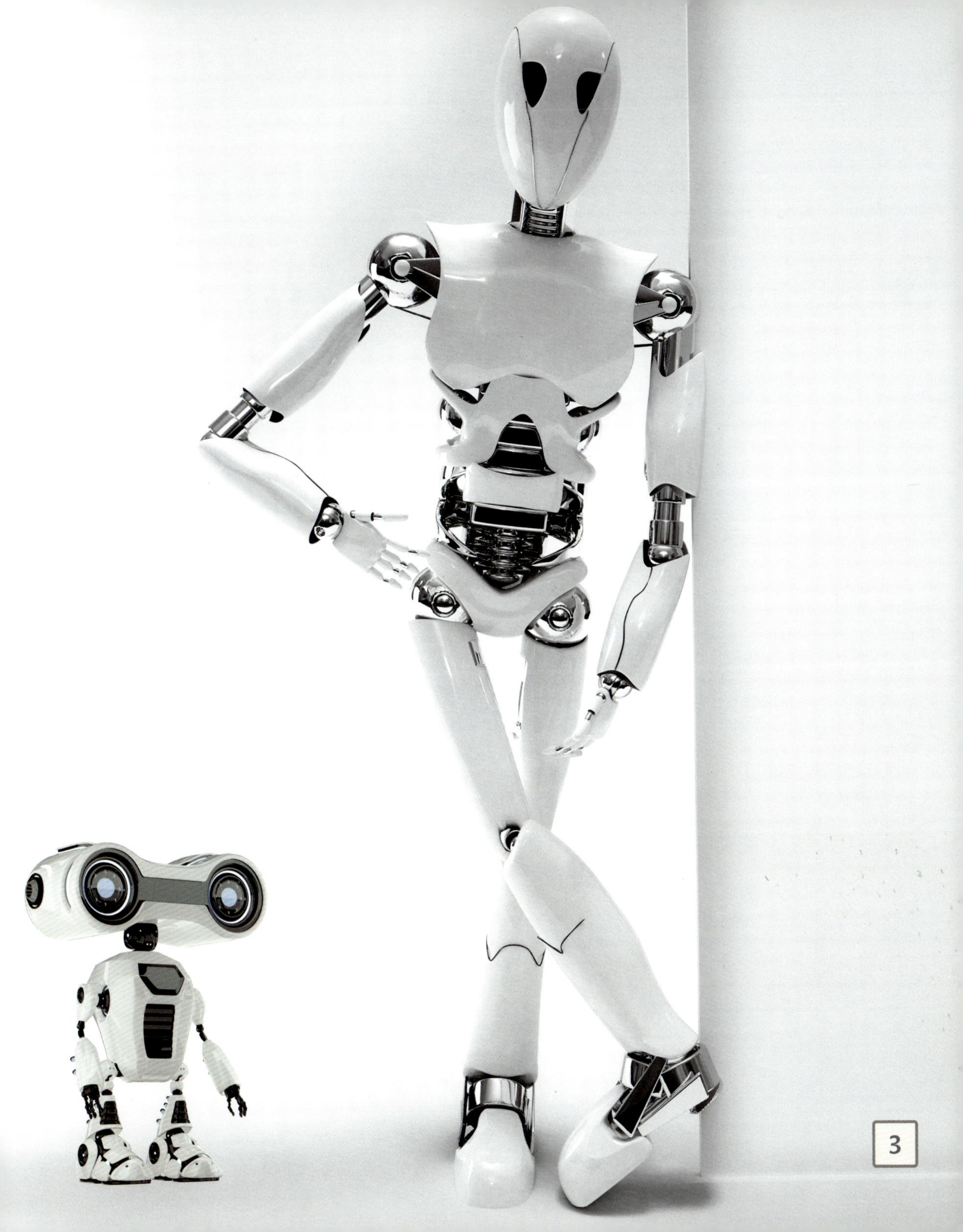

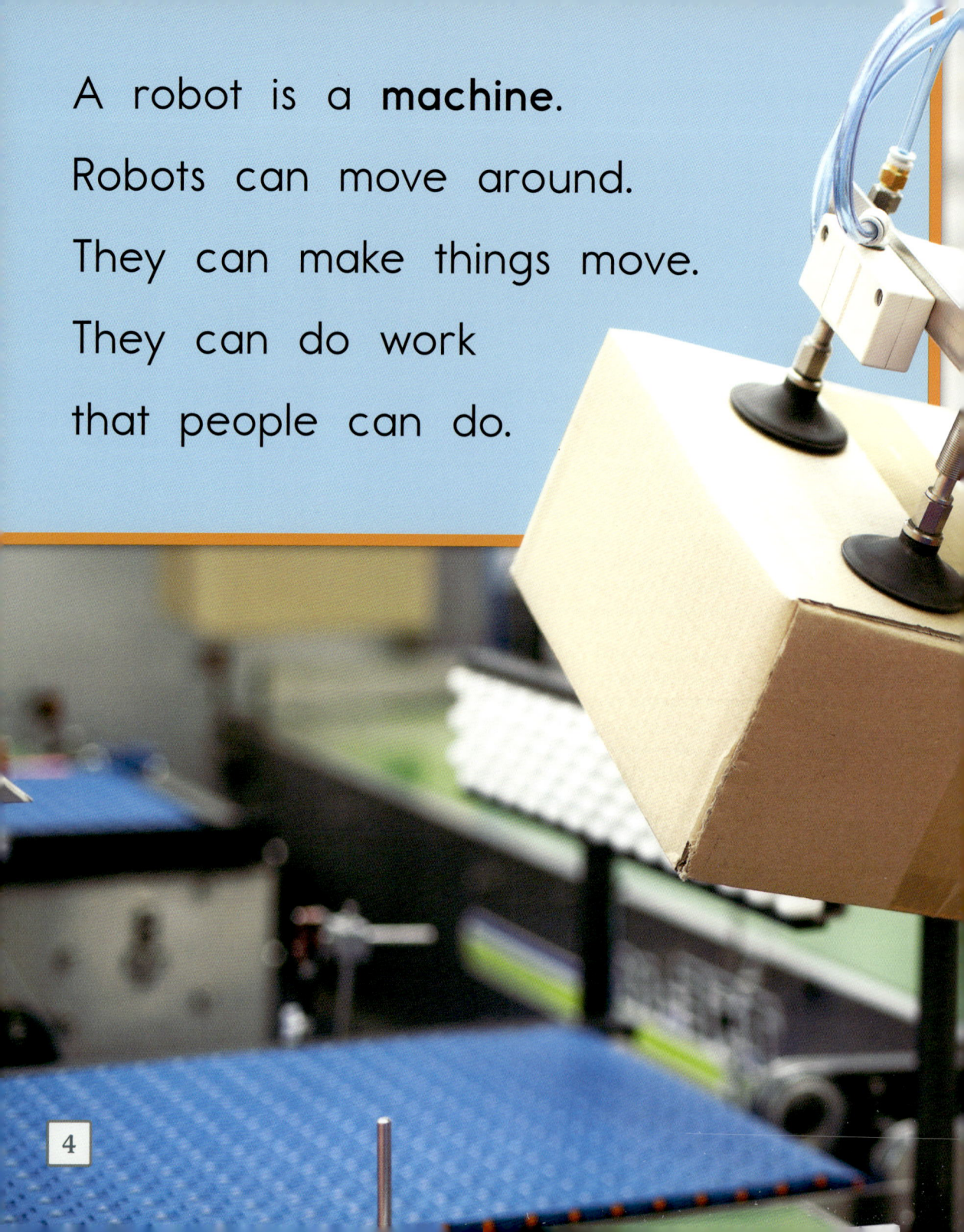

A robot is a **machine**. Robots can move around. They can make things move. They can do work that people can do.

5

WHAT CAN A ROBOT DO?

Robots can do some jobs better than people can. Some jobs need to be done over and over.

People can make mistakes when they do things many times. Robots do not mind doing things over and over. They make fewer mistakes.

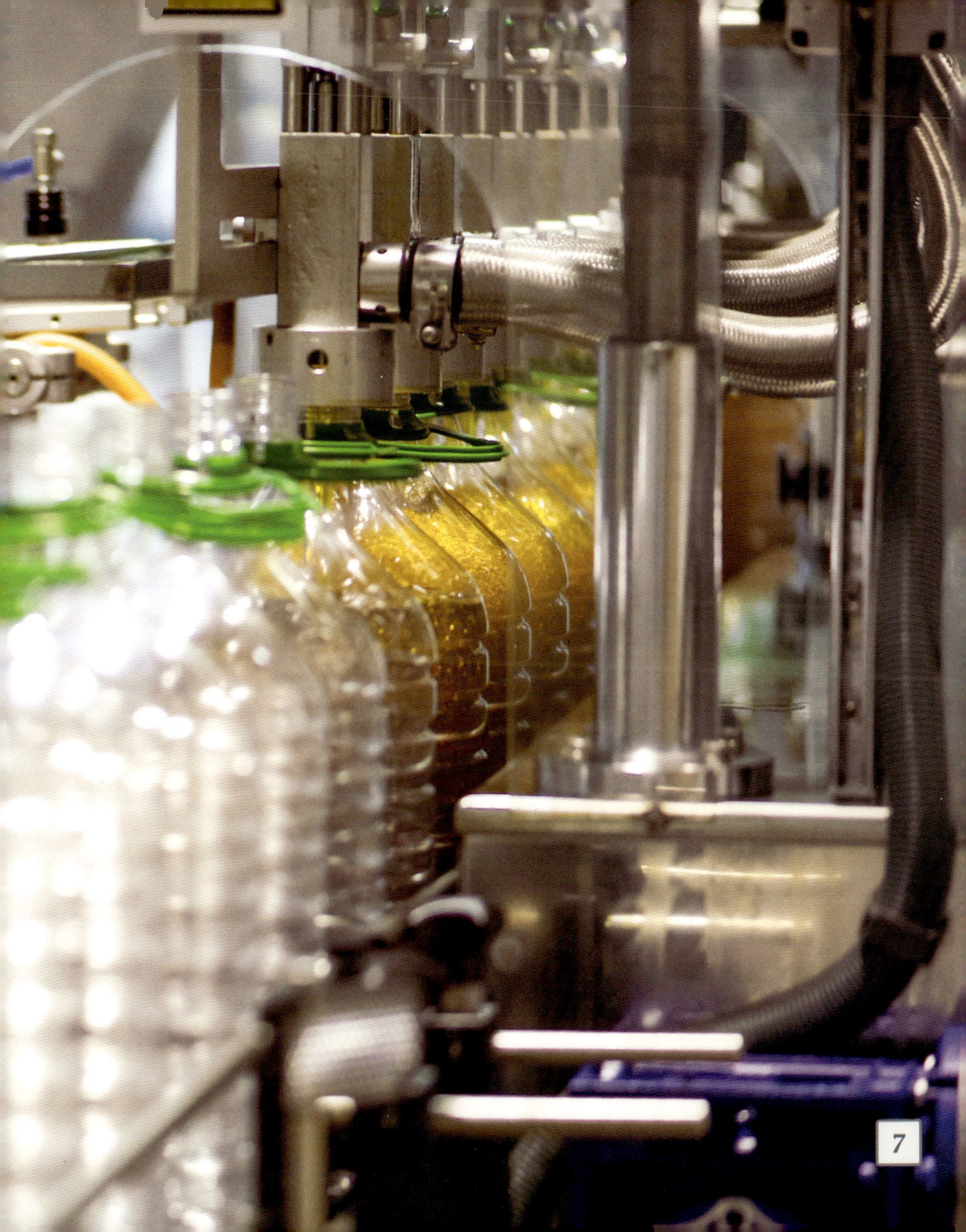

Sometimes robots can do a job that is too dangerous for people.

It is dangerous to go into space. Robots can go into space.

It can be dangerous for people to go into deep water. Robots can go under water and find things.

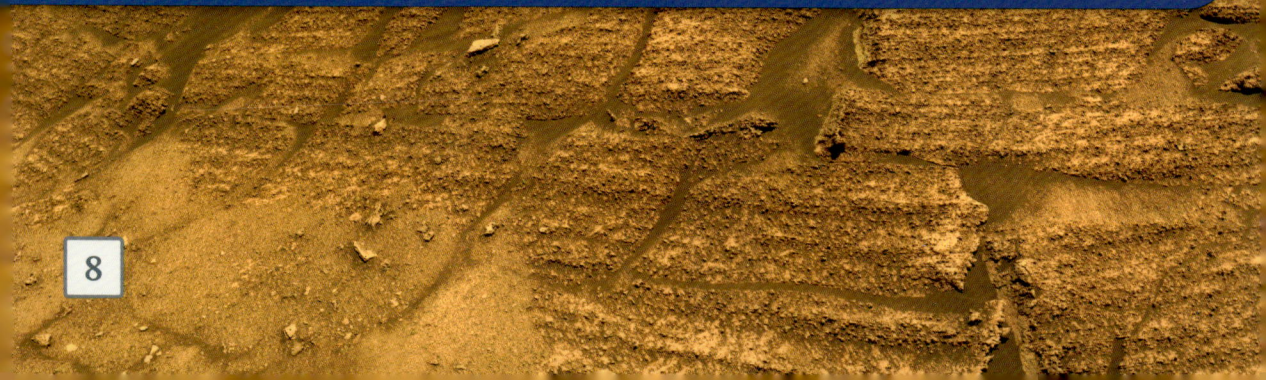

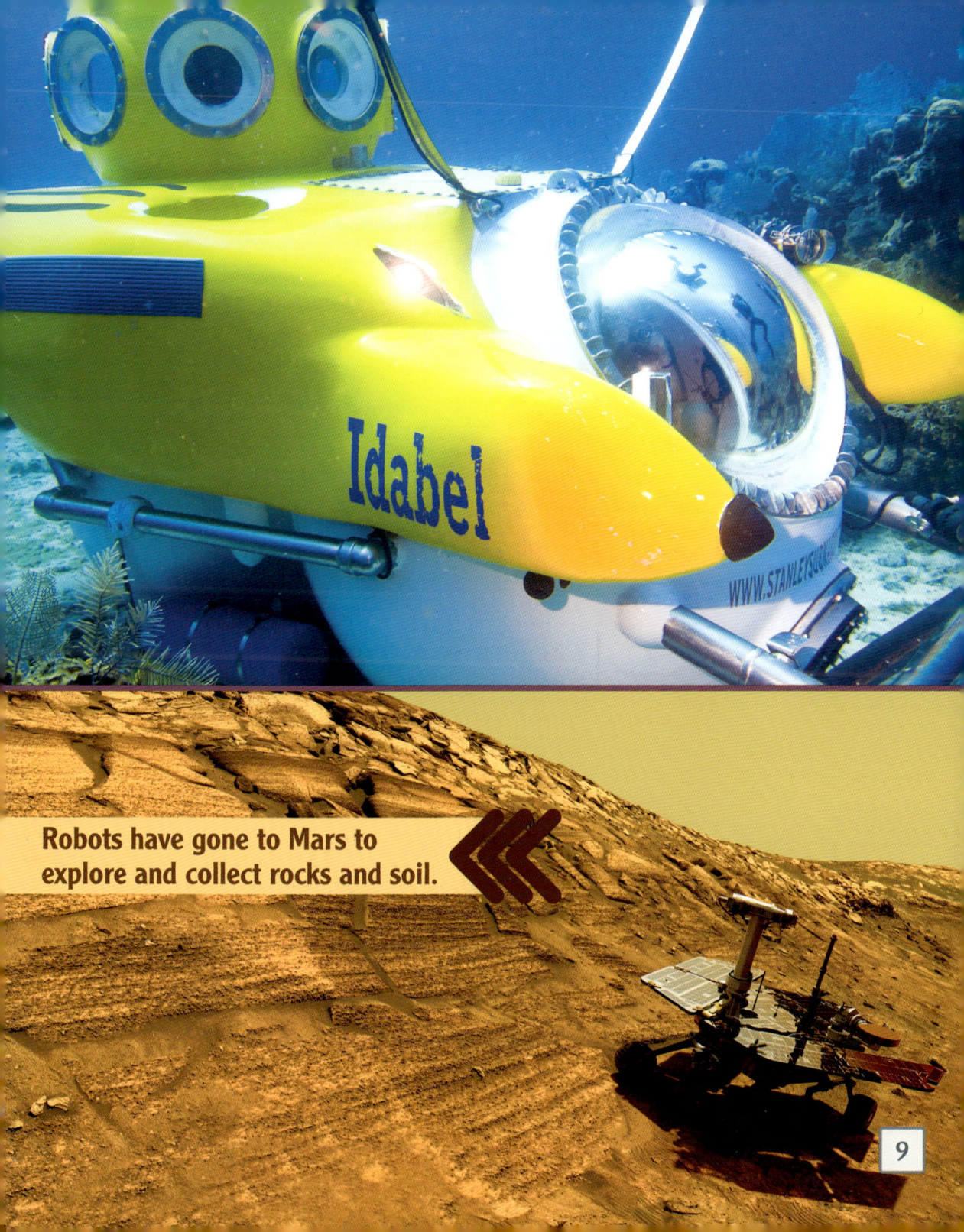

Robots have gone to Mars to explore and collect rocks and soil.

9

Some robots are used to help build cars. They can do some things faster and more carefully than people can.

The first working robot was used in 1961. It helped make cars for the **FORD MOTOR COMPANY**.

ROBOTS AND OPERATIONS

Robots can help doctors, too. A doctor can **control** a robot arm from a computer to perform an **operation**. A robot can make smaller cuts, and makes fewer mistakes.

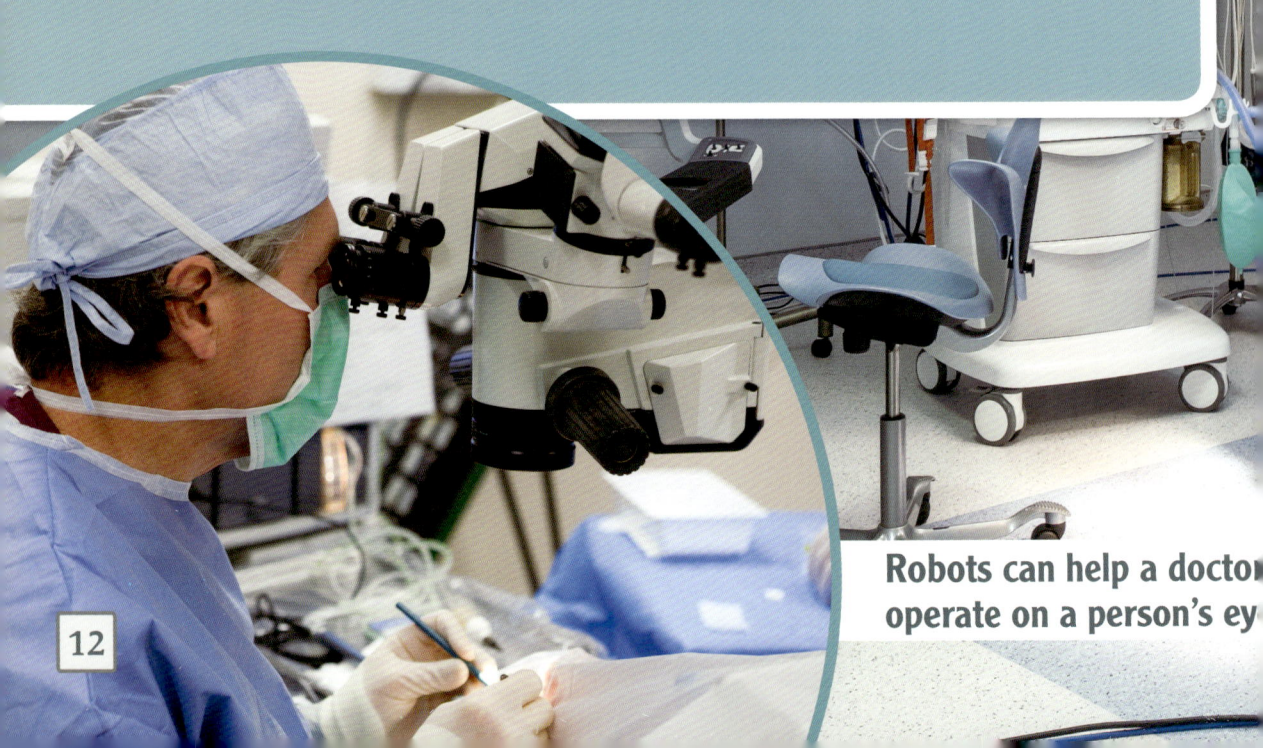

Robots can help a doctor operate on a person's ey[e]

13

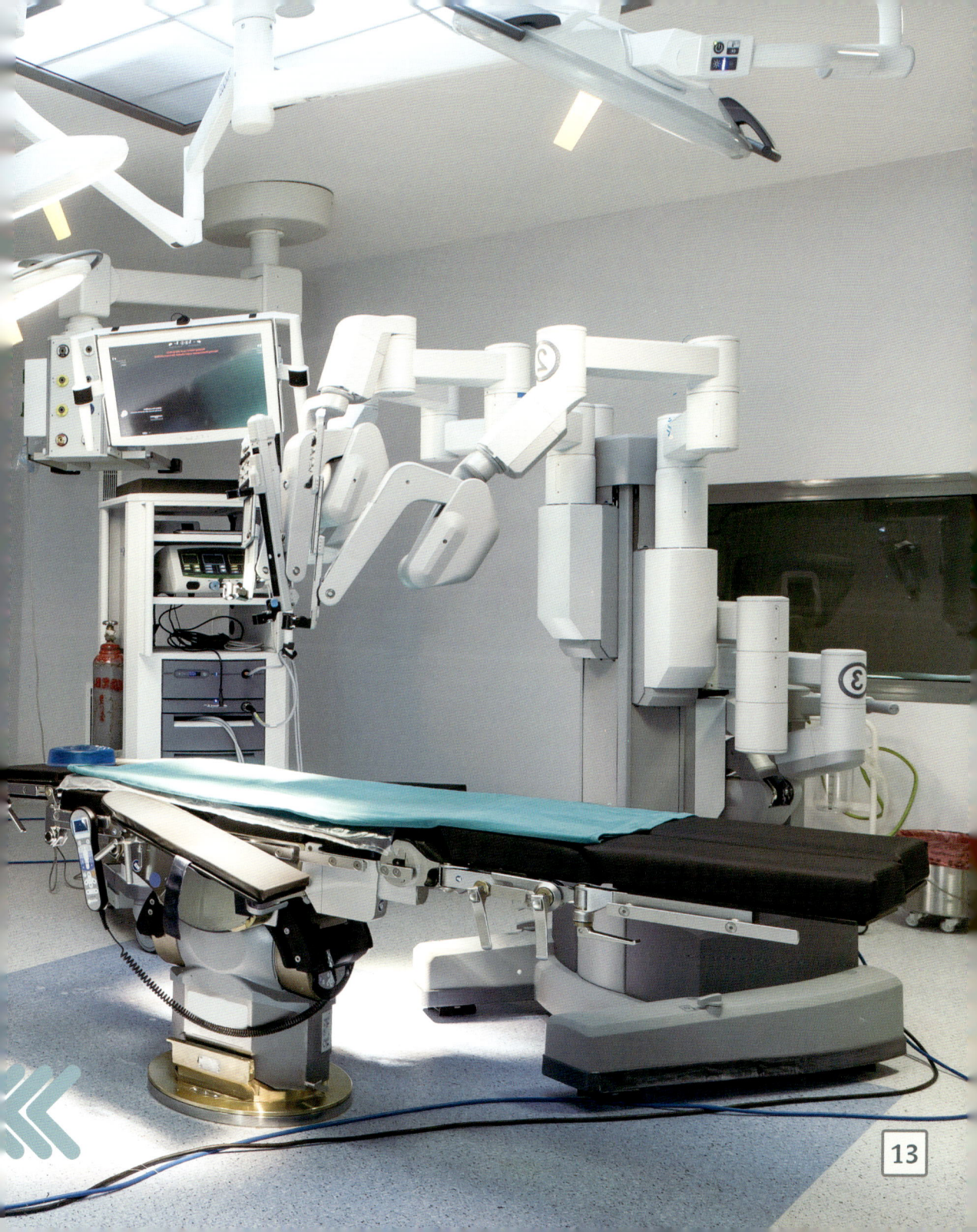

There are things a robot cannot do.

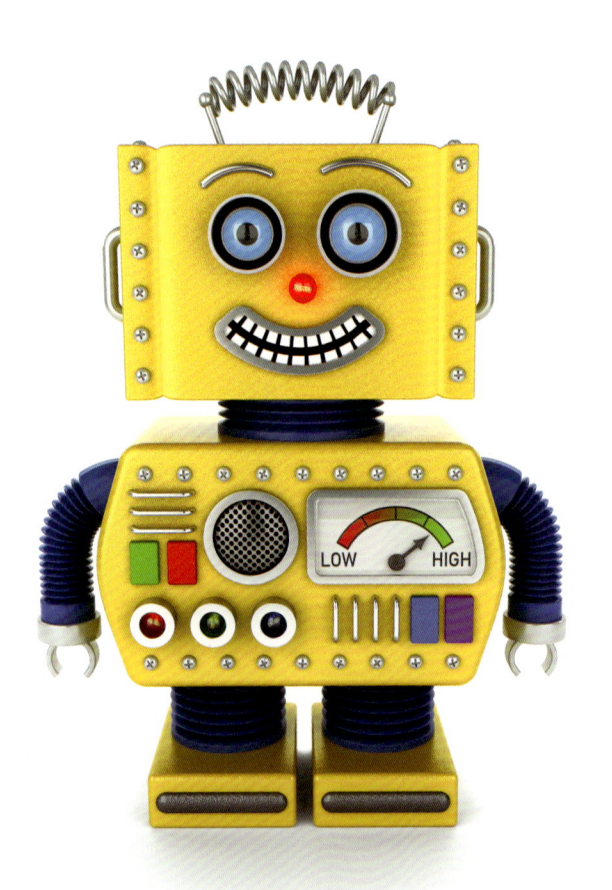

A robot cannot feel **emotions**.

It cannot be happy or sad.

It cannot love or hate.

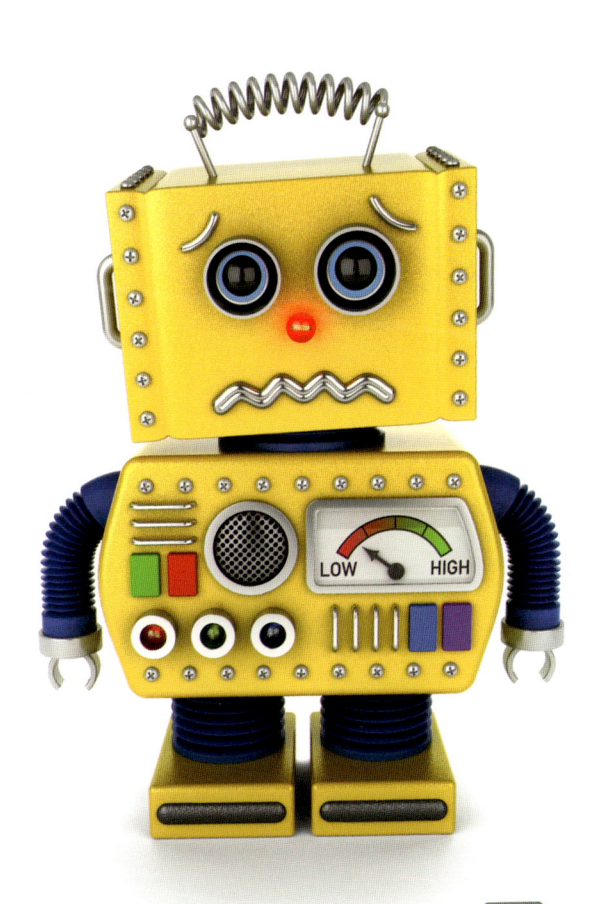

Manufacturing

Service Machines
- ATM
- Vending
- Games
- Car Wash

Robots In Our World

Remote Control
- Cars
- Toys
- Planes
- Helicopters
- Drones
- Vacuum Cleaners

GLOSSARY

control
to have power over

operation
when a doctor cuts into a person's body to fix it

emotions
strong feelings such as joy, sadness, hate, or love

machine
something that has moving parts that, when you give it power, will do work